YOUR KNOWLEDGE HAS VALUE

- We will publish your bachelor's and master's thesis, essays and papers

- Your own eBook and book - sold worldwide in all relevant shops

- Earn money with each sale

Upload your text at www.GRIN.com and publish for free

Beef Cattle Fattening and It's Marketing System. The Case of Damot Pulasa Woreda, Wolaita Zone, Southern Ethiopia

Mesfin Getachew

Bibliographic information published by the German National Library:

The German National Library lists this publication in the National Bibliography; detailed bibliographic data are available on the Internet at http://dnb.dnb.de.

ISBN: 9783346316776
This book is also available as an ebook.

Print and binding: Books on Demand GmbH, Norderstedt, Germany
Printed on acid-free paper from responsible sources.

GRIN web shop: https://www.grin.com/document/966164

Assessment of Beef Cattle Fattening and Marketing System in Case of Damot Pulasa Woreda, Wolaita Zone, Southern Ethiopia

Mesfin Getachew Keshamo

Wolaita Sodo ATVET College, Department of Animal Production, Southern Ethiopia.

Abstract

The study was conducted in the Southern Nation Nationality and People Regional Government in Wolaita Zone to assess beef cattle fattening and marketing system in the case of Damot Pulasa Woreda. In the present study stratified sampling method was used and a total of 120 households were selected for survey study. To collect the data both primary and secondary data source were used and the collected data were analyzed by using descriptive. The study indicated that the main purposes of beef cattle fattening were for 25, (78.1%) start their fattening activity to replace the old oxen after the end of the plowing period and only small proportion 7, (21.9%) of the fatteners were income oriented respectively. The common beef cattle selection criteria were health condition, physical appearance, sex, age are 46.2%, 30.14%, 12.32%, 11.34% respectively. Major feed resources were natural pasture (56.25%), Maize and Enset tuber (31.25%), frushika (10%), grazing (2.5%). Beef cattle fattening constraints were feed shortage (34.4%), lack of capital (25%), market problem (15.6%), disease and parasite (9.375%), lack of extension service (9.375%), and drought (6.25%). The frequency of fattening per year was 21(65.6%) of cattle fatteners in the study area were fattening only one time per year and the rest 7(21.9%) and 4(12.5%) were fattening two times and three times per year respectively. Major beef cattle fattening opportunities were market demand 40.3%, feed resource and water availability 31.7% and comfortable environments 28%. Beef cattle marketing constraints were road problem 43%, unequal demand and surplus 29.2% and market distance 27.8%. Therefore based on the result we endorse that the farmers should be well awarded on beef cattle fattening and marketing system, use improved forage for supplementary feed and should be well informed about market condition and further large scale research should be conducted on the area of beef marketing.

Keywords: Beef cattle fattening, Damot Pulasa, fattened cattle marketing.

Contents

1. INTRODUCTION

Livestock production is an integral part of Ethiopia's agricultural sector and plays a vital role in the national economy. At present, livestock contributes about 20% of the GDP, supporting the livelihoods of 70 % of the population and generating about 11% of annual export earnings (SPS-LMM, 2010). As the country has a large livestock population, which ranks first in Africa and tenth in the world, it has much to gain from the growing global markets for livestock products (SPS-LMM, 2010).

Crop-livestock mixed farming is one of the pre dominant farming systems in the rural communities of south central region. Shortage of land due to population pressure pushed many more farmers either to intensify cropping system or diversify the system using other integrated activities. Cattle fattening is among an integral componential activity (Getahunn, 2008).

At the household level, livestock plays a critical economic and social role in the lives of pastoralists, agro pastoralists, and smallholder farm households. In the case of smallholder mixed farming systems, livestock provides nutritious food, additional emergency and cash income, transportation, farm outputs and inputs, and fuels for cooking food. The government recognizes the importance of livestock in poverty alleviation and has increased its emphasis on modernizing and commercializing the livestock sub-sector in recent years (SPSLMM 2008).

Enhancing the ability of poor smallholder farmers and pastoralists to reach markets, and actively engaging them is one of the most pressing development challenges. Remoteness results in reduced farm-gate prices, returns to labor and capital, and increased input costs. This, in turn, reduces incentives to participate in economic transactions and results in subsistence rather than market-oriented production systems. Sparsely populated rural areas, remoteness from towns and high transport costs are physical barriers in accessing markets (Holloway and Ehui, 2002). For market development, dynamic relationship between demand and supply is a prerequisite, but the smallholder and pastoral livestock production is not market oriented.

Livestock production constraints can be grouped into socio-economic and technical limitations (Mengistu, 2003). Inadequate feed, widespread diseases, poor breeding stock, and inadequate livestock policies with respect to credit, extension, marketing and infrastructure are the major constraints affecting livestock performance in Ethiopia (Desta *et al.*, 2000). However, key constraints to the productivity of livestock in Ethiopian highlands. These include poor nutrition, poor genetic resources in terms of productivity, prevalence of animal diseases, unfavorable socio-economic factors, and lack of livestock policy (Agajie *et al.*, 2002).

2. MATERIALS AND METHODS

2.1. Description of the Study Area

This study was conducted in Damot Pulasa that is one of the woreda in the Southern Nations, Nationalities, and Peoples' Region of Ethiopia, part of the Wolaita Zone. Damot Pulasa is bordered on the east and south by Damot Gale, on the west by the Boloso Sore, and on the north by the Hadiya Zone. Damot Pulasa was separated from Damot Gale woreda. Damot Pulasa is located about 360 km from Addis Abeba and 130 km from Hawassa to the southwest. It is about 25 km from Wolayta Sodo to Northwest.

Based on the 2007 Census conducted by the CSA, this Woreda has a total population of 144,193, of whom 70,736 are men and 73,457 women; 135,239 or 93.79% of its population are rural dwellers.

Damot pulassa is characterized by two agro-ecological zones, Woinadega 96% including 22 Kebele and Dega 4% including 1 kebele. The Woreda was known by large number of livestock, comprising cattle 78,484, sheep 19,163, goat 1,527 poultry 120,190 and equine 2,488. (WZFED, 2018)

There were 23 kebeles in Damot Pulasa woreda from these 6 or six kebeles were selected by using stratified sampling technique due to effects of different agro- ecologies, from these each agro-ecology (Woina-dega and Dega) among those six kebeles, four from Woina dega and two kabele from kola was selected. In the same manner, 20 household were selected from each selected kabele. Then a total of 120 households were used for survey data.

Then the data was collected a beef cattle production, marketing system, price, market out let, major beef marketing in the area.

The data was collected from primary source. The primary data was collected through direct interview stakeholder by preparing questionnaires for the cattle owner, field workers and other expected person.

The secondary data was collected from written documented materials concerning beef cattle fattening and marketing system. Secondary sources kept in Damot Pulasa Woreda Animal and Fishery Resource Office was collected. The secondary data include total livestock population and the document files of the beef cattle fattening and marketing system was used as a source of information.

2.2. Selection of the Participating Households

The participating respondents were drawn from all performers involved in beef cattle fattening and included of producers, rural collectors, wholesalers, retailers and consumers. Damot Pulassa Woreda was purposively selected based on the potential of beef cattle production and comfort ability of climate for cattle feed production and participation of farmers in fattened beef cattle marketing. A total of 6 Kebeles (Peasant Associations) were purposively selected from Damot Pulasa Woreda based on the extent of beef cattle production and marketing. Beef cattle producers and non-producers were identified using stratified sampling technique. Finally the participating farmers were randomly selected.

2.3. Selection of Participating Traders

This study included intermediary producers and actors involved in fattened beef cattle marketing i.e. Wholesalers, assemblers, retailers and consumers. These were selected from the major towns and marketing centers located in Boditi, Sodo, and Shanto towns. These were purposively selected based on their direct involvement in the fattened beef cattle marketing. Traceability and snow ball sampling technique. Retailers were randomly selected from Boditi market and Shanto town.

2.4. Data Collection

The primary data were collected with the use of well-structured and pretested questioners. Both close and open ended (semi- structured) questionnaire were prepared. Enumerators who have college diploma and working as development agents were recruited and trained in data collection. Secondary data were collected through key informants, extension workers, agricultural office workers and fattened beef cattle traders and consumers. Literature and document review were also used to collect secondary data. Both qualitative and quantitative data were collected throughout the study period. Qualitative data collected included characteristics of the actors, fattened beef cattle marketing condition, supporting services along the fattened beef cattle marketing, socio-economic characteristics of the respondents, beef cattle production systems, fattened beef cattle distribution pattern and market outlets. The quantitative data collected included direct and overhead costs incurred by each actor, market margin, percentage share of fattened beef cattle among actors, income from sale of fattened beef cattle, volume of fattened beef cattle production, volume of fattened beef cattle sold and bought, selling and buying price of the fattened beef cattle in unit of measurement, other sources of income, etc. Data on the major constraints and opportunities of fattened beef cattle production and marketing were also collected.

2.5. Data analysis

All the data collected were subjected to appropriate and standard methods of statistical analysis. Descriptive statistics such as frequency, mean, percentage, and standard deviation were used in the process of comparing socio-economic, demographic and institutional characteristics of the respondents. Inferential statistics such as F-test, R^2 and P-value were used to test adequacy of the model and to identify the major challenges and opportunities of potato production and marketing. Marketing margins were analyzed according to (5). Econometric analysis use done with the use of Market supply model suggested by Greene,(2000) using the following

$$Y=X^1\beta +U$$

Where:
Y= quantity of fattened beef cattle supplied to the market
X= a vector of explanatory variables
β = a vector of parameters to be estimated
U= disturbance term

3. RESULTS AND DISCUSSION

3.1. Background information of respondents

Regarding the mean age class of family size in study area, about 0.81±0.11, 1.46±0.12, 1.09±0.12, were males belong to the age class of 0-5, 6-15, and 16-64 years respectively, and 0.78±0.13, 1.15±0.15, and 0.94±0.12, were females belong to the age class of 0-5, 6-15, and 16-64 years respectively (Table 1).
This showed that the largest portions of respondent were within productive age group and this had positive effect in the performance of agricultural activities in the area. On the study area both female and female were participated accordingly the total respondents 64% were males and 36% were females. Most of the farming practiced by the male, like plough, sowing, fattening and house construction while the female were expend their time in home activities like cooking, fetching water, collecting wood and overall family management. The study also showed that from the total respondents, 43.8% were illiterate and the rest 12.5%, 31.2%, 6.20%, and 6.20% had an educational background of basic education, elementary school,

secondary school, and religious education, respectively (Table 2). From this we can recognized the majority of farmers were illiterate.

Table1. Family size and age classes of respondents

		Highland N=60 Mean ±SE	Midland N=60 Mean ±SE	Overall N=120 Mean ±SE	P-value
Age classes	Male (0-5) yrs	0.69 ± 0.12^a	0.94 ± 0.19^a	0.81 ± 0.11	0.280
	Male (6-15) yrs	1.00 ± 0.20^a	1.93 ± 0.26^b	1.46 ± 0.12	0.009
	Male (16-64)yrs	1.31 ± 0.12^a	0.88 ± 0.12^a	1.09 ± 0.12	0.072
	Female (0-5) yrs	0.50 ± 0.16^a	1.06 ± 0.19^b	0.78 ± 0.13	0.032
	Female (6-15) yrs	0.88 ± 0.20^a	1.44 ± 0.22^a	1.15 ± 0.15	0.071
	Female (16-64)yrs	0.94 ± 0.11^a	0.94 ± 0.21^a	0.94 ± 0.12	1.000

Educational status	N	%	N	%	N
Illiterate	30	50.0	24	40.0	54
Basic education	6	10.0	10	16.7	16
Elementary school	14	23.3	10	16.7	24
Secondary school	7	11.7	11	18.3	18
Religious education	3	5.0	5	8.30	8
Total	60	100	60	100	120

As indicated in Table 2, largest land holding size in midland area utilized for cropping and grazing was relatively greater than highland area. This was due to high land area has the largest population whereas on midland area has large portion of land. In high land area for cropping (42%) was higher than grazing land (23.4%). The current study was similar with (Elias *et al.*, 2007) study. In midland, grazing was most common source of feed with limit of the use of crop residue. During wet season, when crop residues are scarce in the highlands, male animals are taken to the midland areas for grazing. In generally, the largest portion of respondents has greater than 0.5 ha of cropping land size i.e. 38%. This showed that many farmers are converting grazing land in to crop lands. This revealed that respondents were utilized most part of land for cropping, thus by products of crop used for beef cattle production.

Table 2: Land holding size

Parameters	Category	Agro-ecology			
		Highland	Midland	Overall	P-value
		N=60	N=60	N=120	
		Mean ±SE	Mean ±SE	Mean ±SE	
Land holding	Homestead	0.16 ± 0.02^a	0.20 ± 0.02^a	0.18 ± 0.01	0.158
	Crop land	1.12 ± 0.08^a	1.97 ± 0.07^b	1.55 ± 0.09	0.000
	Private grazing land	0.39 ± 0.04^a	0.12 ± 0.09^b	0.25 ± 0.06	0.018
	Total private land	**$2.05+0.24^a$**	**2.33 ± 0.20^a**	**2.19 ± 0.09**	**0.146**
Communal grazing land/Kebele/ha		61.12 ± 1.59^a	29.62 ± 1.16^b	45.65 ± 3.03	0.000

3.1 Purpose of fattening

Overall in the study area, the majority of cattle fatteners 25, (78.1%) start their fattening activity to replace the old oxen after the end of the plowing period and only small proportion 7, (21.9%) of the fatteners were profit oriented (Table 3). This was disagreeing with the findings of Daniel Tewodros (2008) of similar study in Borena Zone of southern Ethiopia who reported that the beef cattle are primarily kept for income generation. Despite the contribution of cattle fattening to the household livelihood, the fattening was not adequately market-oriented. This is may be due to the fact that the farming method in the study area was characterized by mixed crop-livestock production system and the livelihood of the household was mainly dependent on crop production and the livestock are being of secondary importance to supply inputs such as draft power, manure, and cash for crop production; almost supported with the report of (Ayele *et al.*, 2003) and (Mohamed-Saleem and Abate, 1995) who reported that livestock are source of agricultural inputs such as draft power and organic fertilizer as a direct contribution for crop production.

Table3. Cattle fattening practices in the study area

Agro ecology

Parameters	Highland(N=60)		Midland(N=60)		Overall(N=120)	
	N	%	N	%	N	%
Purpose of fattening						
Profit	5	31.2	2	12.5	7	21.9
Replace old	11	68.8	14	87.5	25	78.1
Total	16	100	16	100	32	100.0

3.2 Available feed resources over different seasons and feeding systems

The available feed resources and types of grazing systems used for fattening cattle in the study area are presented in the (Table 4). In the study area, the availability of feed resources varied over seasons with respect to quality, quantity and type of feed. The major feed resources available in wet season for fattening cattle in the study area are identified as natural pasture (56.25%), stubble grazing (31.25%), maize and sorghum thinning (6.25%).

On the other hand, the principal feed resources used for fattening cattle during dry seasons were crop residues (43.8%); hay (31.2%) and sweet potato /enset/false banana root (25%).

In general, crop residues and natural pasture were the major feed resources for fattening cattle in the study area and this was almost agree with the report of Tolera *et al.* (2012) who reported that natural pasture and crop residue to be the major feed resources for highlands of Ethiopia. In the highland the majority (87.5 %) of cattle fatteners used swampy grazing land and the rest (12.5%) used open grazing land where as in the midland agro-ecology, all of the cattle fatteners use open grazing land. This is may be associated to the type of grazing land owned communally in each of the study area.

In the study area, different types of grazing systems were employed over different seasons. The major grazing systems used during dry season in highland agro-ecology were herded (81.2%) and tethered (18.8%) were as in the midland agro-ecology herded (62.5%), and tethered (37.5%). The major grazing system employed during wet season in highland were, herded (56.2%), and tethered (43.8%) where as in the midland agro-ecology, herded (68.8%) and tethered (31.2%). This is may be associated with the availability of natural grass and grazing land allocation.

Table 4. Feed resources available for fattening cattle in the two agro ecologies in the study area (rank)

Variables	Category	Agro-ecology				
		Highland N =60	%	Midland N =60	%	P-value
Feed resources in wet season	Natural pasture	10	62.5	9	56.25	
	Taro and sweet potato	4	25.0	5	31.25	0.069
	Maize and sorghum thinning	2	12.5	1	6.25	
	Total	16	100	16	100	
Feed resources in dry season	Crop-residues	10	62.5	5	43.8	
	Enset root	2	12.5	4	25.0	0.018
	Hay	4	25.0	7	31.2	
	Total	16	100	16	100	
Type of communal grazing land	Open grass land	2.0	12.5	16	100.0	0.000
	Swampy grass land	14	87.5	--	--	
	Total	16	100	16	100.0	
Type of grazing system in dry season	Herded	13	81.2	10	62.5	0.079
	Tethered	3	18.8	6	37.5	
	Total	16	100	16	100	
Type of grazing system wet season	Herded	9	56.2	11	68.8	0.060
	Tethered	7	43.8	5	31.2	
	Total	16	100	16	100	

N=number of respondents, x^2=chi-square (level of significant at 5%)   %=percentage

Feeding system of beef cattle

As indicated in Table 5, respondents feed their animals in different feeding system; cut and carry systems (37.5%), both grazing and cut-carrying (28.1%), simple grazing (21.9%) and tethering (12.5%). As respondents responded that the choice of feeding systems were largely depends on the seasonal availability of the feed resources and the amount of family labor required to manage cattle.

Table5. Fattening cattle management practices

Parameters	Highland(N=60)		Midland(N=60)		Overall(N=120)	
	N	%	N	%	N	%
Feeding systems						
Cut-carry	8	50	4	25.0	12	37.5
Tethering	2	12.5	2	12.5	4	12.5
Cut- carry & grazing	4	25.0	5	31.2	9	28.1
Simple grazing	2	12.5	5	31.2	7	21.9
Total	16	100	16	100	32	100

The header row above spans under **Agro-ecology**.

3.3 Watering sources and watering frequency of beef cattle

As regarding with water access for fattening cattle, overall in the study area, there were three types of water sources identified as river (59.4%), spring (25%) and hand dug well (15.5%). The access of water over different season were different when in the rainy season the animals were getting adlibithium accounted for about (59.4%), three times per day (18.8%), once per day (12.5%), two times per day (6.2%), and once per two days (3.1%) depending on the availability of water and interest of the animals to drink. On the other hand during the dry season the beef cattle were getting access to water mostly two times per day (56.2%), and (43.8%) as temporary water sources are dry-off this season and shortage of water occurs. During dry seasons the frequency of drinking water was different between the two agro-ecologies where in highland the majority of cattle fatteners offered two times per day whereas in the midland the larger proportion were giving access to only once per day and this was heavily depends on the climatic condition.

Table6. Water sources and watering frequency

Parameters	Highland(N=60)		Midland(N=60)		Overall(N=120)	
	N	%	N	%	N	%
Water sources for fattening cattle						
Hand dug well	2	12.5	3	18.8	5	15.60
River	10	62.5	9	56.2	19	59.40
Spring	4	25.0	4	25.0	8	25.00
Total	16	100	16	100	32	100.0
Watering during rainy season						
Adlibutum	9	56.2	10	62.5	19	59.4
Three times per day	1	6.2	5	31.2	6	18.8
Two times per day	2	12.5	--	--	2	6.20
Once per day	3	18.8	1	6.2	4	12.5
Once per two days	1	6.2	--	--	1	3.1
Total	16	100	16	100	32	100
Watering during dry season						
Two times per day	13	81.2	5	31.2	18	56.2
Once per day	3	18.8	11	68.8	14	43.8
Total	16	100	16	100	32	100

3.4 Selecting method of cattle for fattening

As showed in Table 7, in the study area beef cattle were selected mostly health condition 34.17% (N=41), physical appearance (30.83%) (N=37), Age (15%) (N=18), Sex (10.83%) (N=13) and color (59.17 %) (N=11). The study agreed with bovine cattle fattening training manual in Hadiya Zone in 2007, beef cattle were selected which has better body condition and medium in age. The physical appearance that to be selected have better body conformation and fast growth rate; both health cows and male animals are preferable. The age beef cattle should not be exceed from 4-6 years old, should be health condition and physical conformation includes rectangular in shape, alert, smooth hide, wide and deep body, big and stand high, healthy and lean. The result of study as revealed in Table 7, conformation and body condition relay up on visual assessment. Besides the health condition of animal is considered in the process, this study also similar with Auriol (1974) who indicated that mortality and morbidity rate are major factors for selection of beef cattle. In the same way, in the current study, most of respondent indicated that major criteria for selection of beef cattle were health condition (34.17%) (N=41) and physical appearance assessment (30.83%) (N=37).

Table 7: Beef cattle selection criteria percentage

Selection criteria	No of respondents	%
Physical appearance	37	30.83
Health condition	41	34.17
Age	18	15
Sex	13	10.83
Color	11	9.17

3.5 Seasonality of beef cattle fattening

As defined in the Table 8, most of the time beef cattle fattening starts from June-September (34.3%) and this was governed by seasonal pattern of feed availability, condition of the environment and market demand. Beef cattle fattening in study area were strategically practical with seasonal feed availability and market demand. The rest of period mentioned by respondents showed scarcity of feed availability. As a result of our findings beef cattle's were fattening throughout the year during dry season. Similar to current result Nega *et al*, (2202) and Amena *et al*, (2007) dry season was typically characterized by shortage of feed.

Table 8: Beef cattle fattening season

Selection criteria	No of respondents	%
January to March	38	31.7
April to June	18	15
June to September	41	34.3
October to December	12	10

3.5 Duration of beef cattle fattening

The result of study revealed in the Table 9, that the length of fattening period varies according to the feed availability, market demand. Most of the respondents in study area feed their cattle consume more feed at starting time. This is due fact that, the animals use for growth and further muscle development and the need of more feed become low at finishing time. Therefore, the new animals are purchased after selling finished once and are fattened turn by turn.

While dealing with the length of feeding of fattening cattle, overall in the study area, the majority (53.1%) and (40.6%) of cattle fatteners were feeding for three and four months respectively and the rest a small proportion only (6.2%) were finish fattening in five months. When calculating these proportions the average length of feeding was estimated to be 120 days. This is somewhat supported with the findings of Shitahun Mulu who reported that the average feeding length of fattening cattle was estimated to be 110 days. The feeding length was associated with the availability of feed and the delivery of fattened cattle to the market when the demand of fattened cattle was high. However; in the highland Agro ecology the greater proportion of cattle fatteners (56.2%) were finish in four months and in contrast the greater proportions of cattle fatteners (75%) in midland agro ecology were end fattening in three months. This was may be due to the variation of feed availability in two agro ecologies as a consequence of climatic variation.

Table 9: Length of beef cattle fattening in months

	Agro-ecology					
	Highland(N=60)		Midland(N=60)		Overall(N=120)	
Parameters	N	%	N	%	N	%
Length of fattening						
Three months	5	31.2	12	75.0	17	53.1
Four months	9	56.2	4	25.0	13	40.6
Five months	2	12.5	--	--	2	6.2
Total	16	100	16	100	32	100

4. BEEF CATTLE HOUSING SYSTEM

The current study (Table.10) indicated that overall in the study area, the two types of housing systems for fattening cattle were identified as separated house constructed for the cattle (56.2%) and separated room together with in family (43.8%). The result showed that separated house was more preferred in midland agro ecology (68.8%) whereas separated room in family house was more preferred in highland agro ecology (56.2%) and this was related to the livestock holding per household.

Regarding the construction of different housing structure from different locally available materials in the study area, mud was almost used for the construction of floor by all of the cattle fatteners and wall was constructed from wood and mud (59.4%), wood (40.6%) and roof was constructed from the grass (Table10). However; the choice of materials was mainly depends up on the availability of materials, the protection of cattle from adverse weather conditions and the management skill of the cattle owners.

Table 10. Beef cattle fattening house construction

	Agro-ecology					
	Highland(N=60)		Midland(N=60)		Overall(N=120)	
Parameters **Types of house**	N	%	N	%	N	%
Fencing	-	-	-	-	-	-
Confined with other Ls	9	56.23	5	31.25	14	43.8
Stall **Cause of housing**	7	43.8	11	68.8	18	56.2
To minimize heat loss	13	21.67	12	22.22	25	20.83
To create warm condition	19	31.67	18	33.33	37	30.83
For close supervision	15	25	13	24.07	28	23.33
To keep cattle from thief	13	21.67	11	20.37	24	20

Marketing of fattening cattle

4.1. Market place

The survey result of current study (Table 11) indicated that the majority of cattle fatteners were purchasing the cattle for fattening purpose from "Shanto" the district market accounted for about (68.8%) followed by "Lera" (21.8%) and "Boditi" the zonal market about (9.40%). This is mainly related to the relative purchasing price, and the proximity of the market to fatteners' destination.

On the other hand; the majority of cattle fatteners were selling the fattened cattle in Shanto market (50.25%) and Boditti (31.00%). This is mainly because in Shanto market there were big traders who took the fattened cattle to other big markets like Addis Ababa, participated and hence the selling price is relatively high.

4.2 Market participants

The survey result of current study (Table 11) revealed that overall in the study area, market participants of fattened cattle in the study area were identified as big traders who supply to big markets like Addis Ababa (25%), local butchers, hotels and restaurants (28.1%) and group consumers (81.25%) found in Shanto, Lera and Boditti markets.

4.3. Demand of fattened cattle

As indicated in (Table 11), about (59.4%) of the respondents were stated that the demand of fattened cattle became high during Muslim festivities like Arafa and followed by Christian festivities like Easter (40.6%). This may be due to the preference of high consumption of cattle meat per house hold during these times relative to other livestock meat whereas the low demand of fattened cattle was commenced during June (59.4%) followed by May (40.6%).

Table11. Marketing of fattening cattle in the area

	Agro-ecology					
Parameters	Highland(N=60		Midland N=60		Overall N=120	
Purchasing place	N	%	N	%	N	%
Shanto	10	62.50	12	75.0	22	68.8
Lera	8	25.00	3	18.8	11	21.8
Boditti	4	12.50	4	25.0	8	18.75
Selling place						
Shanto	9	56.25	5	43.75	14	50.25
Boditi	5	31.25	7	31.25	12	31.00
Market participants						
Local butchers, hotels, and restaurants	6	37.5	3	18.75	9	28.1
Group consumers	2	12.5	11	68.75	13	81.25
Big traders	8	50	2	12.5	10	25
Total	16	100	13	100	32	100
High demand						
During Muslim festivity	10	62.50	9	56.25	19	59.40
During Christian festivity	6	37.50	7	43.75	13	40.60
Total	16	100.0	16	100.0	32	100.0
Low demand						
May	7	43.75	6	37.50	13	40.60
June	9	56.25	10	31.25	19	59.40
Total	16	100.0	16	100.0	32	100.0

%=percentage

4.4. Price of fattening cattle

The survey result of current work (Table 12) showed that overall in the study area, the mean price of cattle before and after fattening was ($6,637\pm1.09$ and $9,562\pm2.28$) birr, ($6,022\pm1.32$ and $8,956\pm1.91$) birr, and ($5,956\pm0.60$ and $9,375$) birr per cattle in Shanto, Lera and Boditti markets respectively and resulted in the gross profit of 2925 birr, 2934 birr, and 3419 birr respectively per fattening cattle which comes from the price and feed margins. The mean purchasing price in Serbo market had a significance difference ($P<0.05$) higher for the cattle from highland agro ecology than midland Agroecology but it had no significance difference ($P>0.05$) lower for the cattle from highland agro ecology than midland Agroecology in Jimma market.

Whereas the mean selling price had a significance difference ($P<0.05$) highest in Boditti market for fattened cattle from highland agro ecology than midland agro ecology and lowest in Lera market between the highland and midland agro ecologies and it had no significance difference ($P>0.05$) higher in Shanto market than Boditti market between the highland and midland agro ecologies. Generally, the mean selling price of fattened cattle from highland agro ecology was higher than from the cattle of midland agro ecology in all markets of study area. This may be due to the higher live-weight change of the fattened cattle in highland agro ecology associated with better feed availability and agro ecological variation.

Table.12. Price of fattening cattle

Parameters	Category	Agro-ecology			P
		Highland	Midland	Overall	value
		N=60	N=60	N=120	
	Markets	Mean $\pm$SE	Mean $\pm$SE	Mean $\pm$SE	
Purchasing price before fattening (birr)/animal	In Lera	$6,168\pm1.15^a$	$5,744\pm0.05^a$	$5,956\pm0.60$	0.355
	In Boditti	$6,856\pm1.59^a$	$6,418\pm1.32^b$	$6,637\pm1.45$	0.043
	In Shanto	$5,975\pm1.09^a$	$6,068\pm2.17^a$	$6,022\pm1.63$	0.730
Selling price after fattening (birr)/animal	In Shanto	$9,394\pm2.67^a$	$8,518\pm1.16^a$	$8,956\pm1.91$	0.005
	In Boditti	$10,094\pm3.27^a$	$9,031\pm2.69^b$	$9,562\pm2.28$	0.018
	In Lera	$9,462\pm3.30^a$	$9,287\pm2.01^a$	$9,375\pm2.65$	0.654

N = number of respondents   SE = Standard Error

a, b = mean with different superscript across rows shows significant difference at (P<0.05)

4.5 Beef cattle marketing and marketing channel

As indicated in Table 13, most of these respondent were sale their cattle after finished for small traders 73.33% (N=44) and some of them sold their cattle directly for butchers 20 % (12). some fatteners were better informed on market price and sold for small traders mostly and butcher, but other/Delala/ price are usually fixed by individual bargaining and depend mainly on supply and demand, which is heavily influenced by the season of the year and the occurrence of religious and cultural festivals (MOA, 1976).

Table 13: Beef cattle marketing and marketing channel

Agro-ecology						
System of marketing	**N**	**%**	**N**	**%**	**N**	**%**
Local butchers, hotels	6	37.50	3	18.75	9	28.10
Group consumers	2	50.0	11	68.75	19	59.40
Small traders	8	12.50	2	12.50	4	12.50
Export marketing	-	-	-	-	-	-
Types of selling season						
Enkutatash holiday	18	30	19	31.67	37	30.83
Muslim holiday	12	20	9	15	21	17.5
Gifaata holiday	24	40	26	43.33	50	41.67
Huge market day	6	10	6	10	12	10
Parameters	Highland		Midland		Overall	
	N=60		N=60		N=120	

System of marketing	N	%	N	%	N	%
Local butchers, hotels	6	37.50	3	18.75	9	28.10
Group consumers	2	50.0	11	68.75	19	59.40
Small traders	8	12.50	2	12.50	4	12.50
Export marketing	-	-	-	-	-	-
Types of selling season						
Enkutatash holiday	18	30	19	31.67	37	30.83
Muslim holiday	12	20	9	15	21	17.5
Gifaata holiday	24	40	26	43.33	50	41.67
Huge market day	6	10	6	10	12	10

4.6 Seasonality of beef cattle fattening

As described in the Table 14, most of the time beef cattle fattening starts from June-September (35%) and this was governed by seasonal pattern of feed availability, condition of the environment and market demand. Beef cattle fattening in study area were strategically practical with seasonal feed availability and market demand. The rest of period mentioned by respondents showed scarcity of feed availability. As a result of our findings beef cattle's were fattening throughout the year during dry season. Similar to current result Nega *et al*, (2202) and Amena *et al*, (2007) dry season was typically characterized by shortage of feed.

Table 14: Beef cattle fattening season

Season of beef cattle fattening	No of respondent	%
January to March	12	10
April to June	27	22.5
June to September	42	35
October to December	39	32.5

4.7 Duration of beef cattle fattening

The result of study revealed that, the length of fattening period varies according to the feed availability, market demand. Most of the respondents in study area feed their cattle consume more feed at starting time. This is due fact that, the animals use for growth and further muscle development and the need of more feed become low at finishing time. Therefore, the new animals are purchased after selling finished once and are fattened turn by turn.

According to the selected respondents the number of cattle finished per cycle varies based on capital stands, feed availability and market demand. Some of respondents said that of beet cattle were fattened 1-3 months 71.66 and for 4-6 month 23.33% which exceeds the maximum length of fattening period to reach targeted fattening level. Furthermore, finished cattle are sold at good price due to maximum consumption of beef during main holidays (Meskel) Easter, Christmas, Enkutatash). Hence, supply, demand and consumption of beef exhibit seasonal trend.

4.8 Beef cattle production opportunities

As illustrated in Table 15, there were beef cattle production different between high lands and mid land areas in study area, market demand 19.2% (N=23) and comfortable environment included climate and weather condition like rain fall, temperature, humidity and the market demand showed consumers demand was high. Some opportunity were also include like feed and water availability 22.5 % (N=27), road access 18.3% (N=22) in order of their importance.

Table 15: Beef cattle production opportunity

No	List of opportunity	Highland No=60	%	Midland No=60	%	Total No=120	%
1	Confortable environments	18	30	17	28.33	35	29.2
2	Feed and water availability	13	21.7	14	23.3	27	22.5
3	Market demand	11	18.3	12	20	23	19.2
4	Road access	12	20	10	16.7	22	18.3
5	Professional support	6	10	7	23.33	13	10.8

4.9. Constraints of cattle fattening

As a result of individual survey supported with focus group discussions and field observations held in both of the study kebeles, the overall in the study area, the major constraints that hindered the performance of cattle fattening activity in both agro-ecologies were mentioned as feed shortage (34.4%), lack of capital (25%), Market problem (15.6%), Disease and parasite (9.375%), Lack of extension service (9.375%), and Drought (6.25%) in order of importance (Table 13). This was somewhat supported with the findings of Shitahun Mulu (2009) in Bure woreda, Amhara region who reported that the major constraints that hindered the Performance of cattle fattening activity were mentioned as feed shortage, lack of capital, shortage of labor, and animal health problem in order of importance.

Table 16. Major constraints for cattle fattening in the study area (rank)

Major constraints	Highland N=60		Midland N=60		Overall N=120	
Constraints	N	%	N	%	N	%
Feed shortage	6	37.5	5	31.25	11	34.40
Disease and parasite	1	6.25	2	12.50	3	9.375
Market problem	2	12.5	3	18.75	5	15.60
Lack of capital	4	25.0	4	25.00	8	25.00
Lack of extension service	2	12.5	1	6.250	3	9.375
Drought	1	6.25	1	6.250	2	6.250
Total	16	100	16	100.0	32	100.0

4.10 Beef cattle marketing constraints

As presented in Table 17, below beef cattle marketing constraints were varied between high land areas and mid land areas, in highland area the road problem 35%, market distance 26.7% and comparatively in mid land areas while the road problem was 17.5%, market distances 15%. Moreover both season's price variance and unequal demand supply were relatively common problem.

Table 17: Types of beef cattle marketing constraints

Types of constraint	Highland N=60	%	Midland N=60	%	Overall N=120	%
Road problem	21	35	21	17.5	42	35
Market distance	16	26.7	18	15	31	25.8
Seasonal price variation	13	21.7	11	9.2	24	20
Unequal demand and surplus	10	16.7	10	16.7	20	16.7

Generally in the study area the major problems of beef cattle marketing and production were road problem 35% and unequal demand and supply 27.33%. The result of present study is similar to Ayele *et al* (2003) who stated that the number of animals accessible in the market is usually greater than the number of demanded, so there is usually excess supply. The study is also agrees with to Holloway and Ehui (2002) who showed that remoteness results in reduced farm date prices to labor and capital and increased input costs. This reduced incentives to participate in economic transactions and results in subsistence rather than market-oriented production system. Sparsely populated rural areas, remoteness from towns and high transport costs are physicals barriers in accessing markets.

5. CONCLUSION

The study showed that the selection criteria for beef cattle were mainly animal age, health condition, sex and physical appearance of the animal. The main purposes of keeping beef cattle were for income generation and consumption. The major feed source for beef cattle in the study area was natural grasses and crop residues. The feeding was mostly by cut and carrying system. Beef cattle fattening season and duration were mainly from June-September and 1-3 months. The common beef cattle production constraints were feed shortage, management, diseases, breed and drought. Marketing constraints were seasonal price variation and unequal demand and supply. Beef cattle marketing were practiced mainly during Meskel holiday and festivals like Gifaata. The length of fattening period varies according to type of feed availability used and market demand. Channel of marketing was mainly done by small traders.

6. RECOMMENDATION

- Capacity building training should be needed for farmers to create awareness about beef cattle fattening and marketing. Empowering the farmers so that they can provide high-quality, sustainable beef cattle production and they should have access to basic production in puts, credit, and market related information.

- Adoption of improved forage by Woreda Animal and Fishery resource office, selection of forage breed, which have better adoption, proper usage of feed and over all managerial activities should be carried carefully.

- The producers should use separate housing for fattening cattle before starting fattening to reduced feed competition by others.

- In generally there is a need from government to provide extension services with the capacity, support and physical means to expose small scale farmers to markets and by so doing, efficiency in production and marketing of cattle to achieve huge profit.

7. REFERENCES

ACTESA. (2011). Botswana Livestock Value Chain Base Line Study.

ACDI/VOCA. (2006b). Baseline Information for Key Indicators. Pastoralist Livelihood Initiative Livestock Marketing Project.

Agajie Tesfaye, Chilot Yirga, Mengistu Alemayehu, Elias Zerfu and Aster Yohannes. (2002). Smallholder Livestock Production Systems and Constraints in the Highlands of North and West Shewa Zones. *In:* Proceedings of the 9th Annual Conference of Ethiopian Society of Animal production (ESAP Held in Addis Ababa, Ethiopia, August 30-31, 2001. pp 49-72.Agricultural Cooperative Development International/ Volunteers in Overseas Cooperative Assistance.

Alemayehu Mengistu. (2002). Forage Production in Ethiopia: A Case Study with Multiplication or Livestock Production. Ethiopian Society of Animal Production (ESAP), Addis Ababa, Ethiopia. 106 pp.

Ayele Solomon, Workalemahu Alemu, Jabar. M.A. and Belachew Hurissa. (2003). Livestock Marketing in Ethiopia. A Review of Structure, Performance and Development Initiatives.Socioeconomic and Policy Research Working Paper 52. International Livestock Research Institute (ILRI), Nirobi, Kenya. 35p.

CSA (Central Statistics Autority). (2003). Statistics Report on Land Utilization. Agricultural Sample Survey Volume iv Addis Ababa, Ethiopia.

CSA (Central Statistical Authority). (2009). *Agricultural Sample Survey, 2008/2009 (2001 EC)*, Report on Livestock and Livestock Characteristics. Statistical Bulletin 446. Addis Ababa: FDRE.

FAO (Food and Agriculture Organization of the United Nations). (2001). Grassland Resource Assessment for Pastoral Systems. FAO Plant Production and Protection Paper No. 162. 150pp

Getahun L. (2008). Productive and Economic performance Ruminant production in production system in High land of Ethiopia .PH.D Dissertation University of Hohenheim Stuttgart Hohenheim Germany.p.21

Fikadu Abate. (1999). Assessment of the Feed Resource Base and the Performance of Draught Oxen of the Traditional Fattening Practice of Smallholder Farmers in the

Eastern Hararghe Highlands. M.Sc. Thesis. Alemaya University of Agriculture, Alemaya. 133pp.

Hurissa, B. (2007). Personal communication.

Livestock Marketing Authority (LMA). (2001). Study on Causes of Cross–Border Illegal Trades in South, Southwest and Eastern Ethiopia, Market Research and Promotion Department, Addis Ababa, Ethiopia.

Mengistu A (2003). Country pasture/forage resources profiles: Ethiopia. Food and Agriculture Organization of the United Nations (FAO).

MoARD (Ministry of Agriculture and Rural Development. (2007). Livestock Master constraints with index of 0.134 that affect cattle Plan Study. MOARD, Addis Ababa, Ethiopia.

MOA (Ministry of Agriculture). (1996a). *Fattening extension manual.* MOA, Animal and Fishery Resource Main Department, FLDP (Fourth Livestock Development Project), Addis Ababa, Ethiopia. 83 pp.

Mohamed A, Ahmed A, Ehui S and Yemesrach A (2004). Dairy Development in Ethiopia. EPTD discussion paper No. 123. International Food Policy Research Institute. Washington, DC. USA.41p.

Mulu S. (2009). Feed resource availability, cattle fattening practices and marketing system in Bure Wereda, Amahara Region, Ethiopia. M.Sc. Thesis. Mekele University. P. 51.

Seyoum, B.; Getnet, A.; Abate, T. (2001). Present Status and Future Direction in Feed Resources and Nutrition Research Targeted for Wheat Based Crop-Livestock Production System in Ethiopia. *In:* Wall, P.C. (Ed): Wheat and Weeds: Food and Feed. Proceedings of the Two Stake Holder Workshops, CIMMYT, Santa Cruz, Bolivia. pp 207-226.

Solomon Desta, (2001). Cattle Population Dynamics in the Southern Ethiopian Rangelands. Utah State University Pastoral risk management Project. Research brief 01-02-PARIMA.

SPS-LMM. (2010). Trade Bulletin Issue I. Focus on Ethiopia's meat and Live Animal Export.

Tolera A, Yami A, Alemu D. (2012). Livestock feed resources in Ethiopia: Challenges, Opportunities and the need for transformation. Ethiopia Animal Feed Industry Association, Addis Ababa, Ethiopia.

Zerihun Hailu. (2002). Land Use Conflicts and Livestock Production in Enset-Livestock Mixed Farming Systems in Bale Highland, South-Eastern Ethiopia. M. Sc. Thesis. The Agricultural University of Norway, Aas. 66 pp.

Zinash Sileshi, Abdule Ebro, J. Stuth, A. Jamma, and J. Nidkummana. (2000). Early Warning System and Coping Strategies for Pastoralists. In Pastoralism and Agropastoralism which way forward? Proceeding 8th annual conference of the Ethiopian Society of Animal Production (ESAP) held in Addis Ababa, Ethiopia. August 24-26, 2000